PowerKids Readers:

Tractors

Joanne Randolph

The Rosen Publishing Group's
PowerKids Press™
New York

For Ryan, with love

Published in 2002 by The Rosen Publishing Group, Inc.
29 East 21st Street, New York, NY 10010

First Edition

Book Design: Michael Donnellan

Photo Credits: pp. 5, 7, 9, 15, 17, 19, 21 © SuperStock; p. 11 © CORBIS/Michael S. Yamashita; p. 13 © CORBIS/George Lepp.

Randolph, Joanne.
Tractors / Joanne Randolph.— 1st ed.
p. cm. — (Earth movers)
ISBN 0-8239-6028-5
1. Tractors—Juvenile literature. [1. Tractors.] I. Title.
TL233.15 .R35 2002
629.225'2-dc21

00-013009

Manufactured in the United States of America

Contents

This is a tractor.

CASE

Tractors are big and strong. Tractors are used to pull things.

Tractors are used on farms.

The farmer drives the tractor. He uses it for many jobs around the farm.

M F
185

This tractor is pulling
a plow. The plow
breaks up the ground.

This tractor moves a bale of hay. A bale is a bundle, or ball, of hay.

INTERNATIONAL
1086

This tractor is pulling
many bales of hay.
Tractors can move small
loads or large loads.

2290
CASE
DRV3

This tractor is spraying the fields. The spray helps to keep the plants healthy.

Tractors are busy machines. They do a lot of work on the farm.

JOHN DEERE
4320
DIESEL
JOHN DEERE

Words to Know

bales

plow

tractor

Here are more books to read about tractors:

Cutaway Farm Machines: Look Inside Machines to See How They Work
By Jon Richards
Copper Beach Books

Trucks, Tractors, and Cranes (How Science Works)
By Bryson Gore
Copper Beach Books

To learn more about tractors, check out this Web site:

www.howstuffworks.com/cat-tractor.html

Index

Word Count: 105

Note to Librarians, Teachers, and Parents

PowerKids Readers are specially designed to help emergent and beginning readers build their skills in reading for information. Simple vocabulary and concepts are paired with photographs of real kids in real-life situations or stunning, detailed images from the natural world around them. Readers will respond to written language by linking meaning with their own everyday experiences and observations. Sentences are short and simple, employing a basic vocabulary of sight words, as well as new words that describe objects or processes that take place in the natural world. Large type, clean design, and photographs corresponding directly to the text all help children to decipher meaning. Features such as a contents page, picture glossary, and index help children get the most out of PowerKids Readers. They also introduce children to the basic elements of a book, which they will encounter in their future reading experiences. Lists of related books and Web sites encourage kids to explore other sources and to continue the process of learning.